Busy Bugs

ANTS

By Bray Jacobson

Please visit our website, www.garethstevens.com. For a free color catalog of all our high-quality books, call toll free 1-800-542-2595 or fax 1-877-542-2596.

Library of Congress Cataloging-in-Publication Data

Names: Jacobson, Bray, author.
Title: Ants / Bray Jacobson.
Description: New York : Gareth Stevens Publishing, [2022] | Series: Busy bugs | Includes index.
Identifiers: LCCN 2020006184 | ISBN 9781538263174 (paperback) | ISBN 9781538263181 (6 Pack) | ISBN 9781538263198 (library binding) | ISBN 9781538263204 (ebook)
Subjects: LCSH: Ants–Juvenile literature.
Classification: LCC QL568.F7 J34 2022 | DDC 595.79/6–dc23
LC record available at https://lccn.loc.gov/2020006184

First Edition

Published in 2022 by
Gareth Stevens Publishing
111 East 14th Street, Suite 349
New York, NY 10003

Editor: Kristen Nelson
Designer: Katelyn E. Reynolds

Photo credits: Cover, p. 1 Danita Delimont/ Gallo Images / Getty Images Plus; p. 5 Luis Diaz Devesa/ Moment/Getty Images; p. 7 arlindo71/E+/Getty Images; pp. 9, 24 (colony) Bryan Jones / EyeEm/Getty Images; pp. 11, 24 (nest) Philippe Intraligi / EyeEm/Getty Images; p. 13 Giuseppe Zanoni/Moment/ Getty Images; p. 15 xu wu/Moment/Getty Images; p. 17 Stephane Pattiera / EyeEm/Getty Images; p. 19 ABimagestudio / iStock / Getty Images Plus; pp. 21, 24 (queen) Kaan Sezer/ iStock / Getty Images Plus; p. 23 TheDman/ iStock / Getty Images Plus.

Printed in the United States of America

CPSIA compliance information: Batch #CS22GS: For further information contact Gareth Stevens, New York, New York at 1-800-542-2595.

Contents

Ants are small bugs.

They have three body parts.

They live in groups.
These are colonies.

Ants build nests.

They may be made of soil. They may be made of wood.

Worker ants clean
the nest.
They keep the
colony safe.

Workers find food.

Ants eat plants
and seeds.
They eat
almost anything!

Ant queens lay eggs.

Babies take weeks to grow up.

Words to Know

colony

nest

queen

Index